IMAGES
of America

BAKERSFIELD

IMAGES
of America

BAKERSFIELD

Robert E. Price
Foreword by Mike McCoy

ARCADIA
PUBLISHING

Published by Arcadia Publishing
Charleston, South Carolina

Library of Congress Control Number: 2022950939

For all general information, please contact Arcadia Publishing:
Telephone 843-853-2070
Fax 843-853-0044
E-mail sales@arcadiapublishing.com
For customer service and orders:
Toll-Free 1-888-313-2665

Visit us on the Internet at www.arcadiapublishing.com

*To all those in Bakersfield who find value in the
preservation and celebration of local history.*

CONTENTS

FOREWORD

Bakersfield came into existence in the 1860s, after the gold euphoria of the previous decade had begun to wear off and the serious business of building the West was underway. Central California's expansive agricultural industry grew with the railroads and the taming of the region's waterways. The discovery of oil created a second economic pillar and put a final stamp on Bakersfield's identity.

It was to this Bakersfield that I, the son of Texas immigrants, was born. Like many of my friends, my folks were farmers who also worked in the oil fields. In school, we did not have a strong sense of social castes. Bakersfield had a healthy and pleasant lack of sophistication.

Moving away to attend a large university meant serious culture shock. When I told my new friends I was from Bakersfield, I almost felt like an exchange student. My college summers were spent 60 feet in the air on an oil rig. Theirs were spent as counselors at tennis camps or working at sailing shops in the San Juans.

The breakthrough came when two friends stole a "Bakersfield City Limits" sign from the interstate and installed it in their dormitory. Suddenly, I didn't need to apologize for Bakersfield at fraternity parties; I felt a renewed pride in my hometown, and my Berkeley friends sensed it. I brought them home on weekends and took them to honky-tonks, to the frothy Kern River, and to raucous Basque restaurants. I culture-shocked them right back.

At the Kern County Museum, I continue to culture-shock people into an appreciation of this unique region. As a journalist, Robert E. Price has been doing much the same thing. Images of America: *Bakersfield* is another achievement in that mission.

I hope, after reading this book, folks who are passing through Bakersfield on the interstate, or flying over it, will have a new appreciation for my hometown. It is not just flyover territory, but a place rich with strong people whose stories need to be shared.

Enjoy this book, then come on by for a visit. The first round is on me.

—Mike McCoy
Executive Director, Kern County Museum

ACKNOWLEDGMENTS

Mike McCoy has done a praiseworthy job in his five years as executive director of the Kern County Museum, rejecting inherited remnants of the mundane and adding elements of entertainment to the museum's important role as a repository of local history. One of his best moves was hiring Rachel Hads as curator. Her keen eye and astounding patience were vital in the selection and preparation of the photographs for his book. Cindy Dodd deserves my gratitude as well. After rediscovering and delighting in the work of her photographer father, Bernie Dodd, whose camera chronicled his teen and 20-something years in Bakersfield in the 1940s and 1950s, she allowed me to share them with the world. Thanks also to Jim Shaw of the Buck Owens Private Foundation, Bakersfield High School archiving teacher Ken Hooper, and Ashleigh Meyers and the Arvin-Edison Water Storage District.

INTRODUCTION

Traveling north from Los Angeles, angling away from the Pacific Ocean on Interstate 5, drivers enter a craggy, mostly treeless mountain range known as the Sierra Pelona. Forty-five minutes out, travelers reach the Tejon Pass, 4,100 feet above sea level, then begin a gradual descent toward the long spine of the California interior. At the one-hour mark, the road begins to descend sharply, and suddenly, from behind the last rocky outcropping, a wide, flat panorama comes into view.

On a rare clear day, northbound travelers may notice a distant, indistinct blot of gray at the center of the gauzy valley horizon. That blot, surrounded by a half-million acres of brown and gold farmland, is Bakersfield, the southernmost city of consequence in California's vast San Joaquin Valley. If the Central Valley, as it is more commonly known, were a state unto itself, it would be the nation's 41st largest at 27,478 square miles, somewhere between West Virginia and South Carolina, and 27th in population at 4.3 million, a little more than Oregon.

Except for the speed and means of travel, and the absence of defined squares of cultivated acreage, that passage through the Sierra Pelona is largely how the journey must have been for Edward Fitzgerald Beale in 1865. Twenty-two years before, Beale, along with his friend Kit Carson, had been a hero of the Mexican-American War's 1846 Battle of San Pasqual. Two years after that, on the orders of Pres. James K. Polk, Beale had set the California Gold Rush into motion by bringing to Washington physical proof of the land's subterranean bounty. He then spent a decade on an assortment of government infrastructure development assignments across the West for which, among other undertakings, his men surveyed and cut new roads through the untamed, barren outback, including what would essentially become Interstate 40.

Now, in the 1860s, from his residence at the former US Army base he had helped establish at Fort Tejon as a bulwark against raids by Paiutes and other tribes, Beale made occasional treks across the Sierra Pelona to a settlement 40 miles south: a way station at the confluence of meandering swamps known as Kern Island.

The most prominent settler at Kern Island was a tall man of soldierly bearing and distinguished manner by the name of Col. Thomas Baker. The scion of a well-respected English family, Baker—who obtained his military rank, perhaps honorary, in the Ohio militia—had taken up residence at Kern Island in 1863. He had initially been lured west by the Gold Rush but eventually decided he was better suited to public administration and politics. He settled in the town of Visalia and won election to the state senate, but suspicions associated with his purported Civil War loyalties prompted him to change his mind a second time. He picked up and moved 100 miles south, purchased 160 acres and a log cabin from pioneer Christian Bohna, and set himself to the task of reclaiming swampland.

Native tribes lived in the southern San Joaquin Valley for thousands of years before white men arrived. The first of these interlopers was Spanish soldier and explorer Don Pedro Fages, who claimed the region for Spain in 1769. In 1776, Franciscan friar Francisco Garcés, or Father Garcés as he is remembered today, passed through the place that eventually became Bakersfield, and by

1794, the region was part of the vast territory known as New Spain. Native tribes controlled what is now Kern County for centuries prior, and archaeological sites in southern and central California linked to the Chumash and other tribes collectively known as the Yokuts date back 15,000 years.

English-speaking settlers gradually began to move in after the 1848 Treaty of Guadalupe Hidalgo transferred to the United States the present-day states of California, Nevada, Utah, and most of New Mexico. In 1866, the California legislature created Kern County by merging the southern portion of Tulare County to the north with land from Los Angeles and San Bernardino Counties to the south and east. It was to that newly created, New Jersey–sized jurisdiction that pioneers like Col. Thomas Baker established Civil War–era homesteads.

Thanks to Colonel Baker's efforts, the settlement at Kern Island grew quickly in stature as a marketplace and supply center. Assisted by his young third wife, the previously widowed Ellen M. Whalen, whom he married in 1857, along with daughters Mary and Nellie and son Thomas, Baker cultivated a welcoming sense of place that travelers came to know as Baker's Field. The town adopted Bakersfield as its formal name in 1869, the year a post office was established.

Beale, who had purchased Fort Tejon from the federal government in 1865 and ran it for another six years as the private Tejon Ranch, returned to Washington, DC, in 1871 and, at the urging of his friend Pres. Ulysses S. Grant, served as ambassador to Austria-Hungary for two years, from 1876 to 1877. He then retired to Decatur House, his estate one block north of the White House.

Baker remained a strong force in the growth and development of the town that bore his name and remained at the center of its government and cultural life, such as it was, until his death from typhoid fever in 1872. The following year, Bakersfield was incorporated as a municipality, and in the year after that, 1874, it became the county seat, displacing the Sierra Nevada gold-mining town of Havilah.

Bakersfield owes Beale its gratitude not so much for his personal contributions as for his role in the introduction of his son Truxtun Beale to the emerging frontier town. Truxtun Beale gifted Bakersfield with its first free public library, the Beale Memorial, which, now in its fourth location, still operates today; the Beale Memorial Clock Tower, which was essentially relocated from the center of one of the city's busiest intersections to the Kern County Museum; and the city's first community park, Beale Park, which remains a well-maintained and popular gathering place. In recognition, the city renamed Railroad Avenue—its main east-west boulevard—Truxtun Avenue. Nearly all of the city's municipal, county, and federal offices are on Truxtun Avenue, which appropriately transitions into Beale Avenue at its eastern terminus. Truxtun Beale, who inherited Tejon Ranch from his father in 1893, was never a full-time resident of Bakersfield, but his contributions created a sense of place and pride in a young city that was at the time only starting to consider its place in the world.

The other important family, of course, is that of the city's cofounder and namesake, Thomas Baker, who did not live long enough to see much of its promise come to fruition. His widow, Ellen Baker Tracy, lived a long and generous public life, dedicating much of her time, attention, and wealth to the welfare of children. She founded the area's first school, where she herself was a teacher, and coordinated the development of its first orphanage, which occupied an entire downtown block—in various business capacities—for more than a century.

Her third husband, Ferdinand Tracy, was a tremendously successful grower who employed hundreds and built a farming operation that still thrives today. Ellen Tracy's son, also Thomas Baker, was the city's first constable following its permanent incorporation in 1898. (The city had previously been incorporated and then disincorporated.)

The most important and influential member of the extended Baker family, however, was a non-blood relative, Henry Jastro, a German immigrant who married Baker's stepdaughter after Baker's death. Jastro had several business interests, but most noteworthy was his association with the Kern County Land Company, which owned more than one million acres across the West.

The land company, founded by James Haggin and Lloyd Tevis in 1890, was purchased in 1970 by Houston-based Tenneco West, whose business portfolio has included automotive parts, natural gas transmission, and land development. Tenneco West in turn sold its Bakersfield holdings in 1987 to Castle & Cooke, the parent of Dole Foods Inc., which over its corporate lifetime evolved from pineapple farming to land development. Castle & Cooke became the single most important force in the development of west Bakersfield, building the huge and ever-expanding Seven Oaks residential community as well as the Marketplace commercial center, among other enterprises. That evolution, which brought Bakersfield to prominence as California's ninth-largest municipality—right between the major-league cities of Oakland and Anaheim—might not have happened without Henry Jastro.

Other influences have defined and refined the city's culture and economy. The Dust Bowl migration of the 1930s and 1940s brought many thousands of mostly poor, working families from the Southwest, Midwest, and western Deep South to California, and especially to Kern County—between 150,000 and 400,000, according to various sources and criteria. These Okies, as they called themselves, brought their own patterns of speech, ways of worship, and preferences for food and music with them, and much of it caught on with the broader population.

One of those preferences evolved into the Bakersfield Sound, a subgenre of country music that combines rockabilly and Western swing and produced the two top country-music recording artists of the 1960s, Buck Owens and Merle Haggard. Country music proved to be a good fit with the region's other economic powerhouse beside agriculture, the oil industry, which gained a foothold in the 1910s. Oil helped drive America's love affair with the automobile and created blue-collar prosperity for Kern County families for more than a century.

Over the course of that century, Bakersfield endured triumph and tribulation, both natural and man-made, fleeting and enduring. The city assumed leadership roles in energy and agriculture, music and letters. It survived fire, earthquake, dust, and drought; it wallowed in the darkest depths of corruption and ascended to the highest ideals of leadership.

I have been telling the story of Bakersfield by way of various media, especially newspapers and television, for more than 30 years. This volume, which is built around vintage photographs, represents a somewhat different way of telling these stories. I hope readers will find it every bit as compelling as anything I have shared with them before.

This, in words and pictures, is the story of Bakersfield.

One

BEGINNINGS

Stretching from the Pacific to the Sierra Nevada, the San Joaquin Valley, looking north from its southern tip near Bakersfield, is a fertile expanse with few equals. The region was transformed into year-round cropland through the manipulation of water flow, a transformation evident nowhere more so than Baker's Field, where white settlers tamed a network of swamps from the Kern River and built an agricultural dynamo. (Courtesy of Kern County Museum.)

Edward Fitzgerald Beale settled in what would eventually become Kern County by an accident of history. Born in Washington, DC, in 1822, he graduated from the Philadelphia Naval School in 1842 and traveled to California on the frigate *Congress*. His intrepid battlefield heroics in the Mexican-American War furthered the government's trust in his resourcefulness, leading to a postwar leadership assignment at Fort Tejon. (Courtesy of Kern County Museum.)

Thomas Baker was born in Zanesville, Ohio, in 1810. After serving at 19 as a colonel in the Ohio militia and later studying law, he migrated to Illinois and later to Iowa. When gold was discovered in California in 1848 he moved west, arriving in Tulare County in 1852, where he helped found Visalia, and then 100 miles south to Kern Island in 1863. (Courtesy of Kern County Museum.)

Within two years, Baker had established himself as the leading citizen of Kern Island, which travelers had come to know as Baker's Field. When a post office was established there in 1869, it took the name Bakersfield. In 1857, the legislature gave a group of land speculators the authority to reclaim huge swaths of swampland, and Baker associated himself with the enterprise. By 1866, he controlled 87,120 acres. (Courtesy of Kern County Museum.)

At the urging of Edward Fitzgerald Beale, Fort Tejon was established by the US Army in 1854 to protect and control the Native Americans on the Sebastian Indian Reservation and to protect both them and white settlers from raids. Fort Tejon was abandoned in 1864, and in 1866, Beale purchased the Mexican land grants that now comprise the 270,000-acre Tejon Ranch. (Courtesy of Kern County Museum.)

The mountain village of Havilah was the region's first boomtown. The Gold Rush brought hundreds and perhaps thousands of miners—and an inordinate number of Southern sympathizers such as adventurer-swindler Asbury Harpending—to what would become Kern County. By 1880, Havilah, pictured here in 1910, was a shadow of its former self, with the county seat and the town's population having moved to Bakersfield. (Courtesy of Kern County Museum.)

In 1876, the community's first major, distinctive landmark was built—the original Kern County Courthouse. The handsome edifice, near the dusty intersection of Chester Avenue and what was then called Railroad Avenue, housed all local government offices, serving a city population of about 800 and a county population of 5,000. (Courtesy of Kern County Museum.)

A second generation of settler-founders took the place of Edward Beale and Thomas Baker, including Baker's widow, Ellen Alverson Whalen Baker, a hearty woman who had already survived much by the time of his death. Widowed at 15 and again at 39, she married prominent rancher Ferdinand Tracy in 1875 and retained her status as Bakersfield's first lady. (Courtesy of Kern County Museum.)

Ellen Baker Tracy's passion was the welfare of children. As a girl, she herself had endured much, traveling across the Donner Pass into California in an ox-pulled wagon. She was married as a teen and left to care for two young daughters upon her first husband's death. With the difficulties of pioneer children in mind, she opened the town's first school—and served as teacher. (Courtesy of Kern County Museum.)

Ferdinand Tracy died in 1908, widowing Ellen Baker Tracy for a third time. Eleven months later, at her urging, Bakersfield residents raised $6,000 for construction of the city's first orphanage. The Kern County Children's Shelter opened in 1909 on two acres donated by the Tracys. Ellen Baker Tracy died in 1924. The orphanage, later a series of restaurants, was razed in 2021. (Courtesy of Kern County Museum.)

Truxtun Beale, the only son of Mary and Edward Fitzgerald Beale, was named for his grandfather Commodore Thomas Truxtun. Beale followed in his father's footsteps and served in diplomatic assignments to Iran, Serbia, Romania, and Greece. Upon his father's death in 1893, however, and having inherited the Tejon Ranch, Beale returned to California and managed its 270,000 acres for the next 13 years. (Courtesy of Kern County Museum.)

Between 1886 and 1919, philanthropist Andrew Carnegie donated $40 million for the construction of 1,679 new libraries in communities across America. Bakersfield, perhaps too small and too young, was not among them. Truxtun Beale, however, answered the call, financing the city's first free public library at Seventeenth Street and Chester Avenue in 1900. He named it the Beale Memorial Library in honor of his father. (Courtesy of Kern County Museum.)

In 1903, Railroad Avenue was renamed Truxtun Avenue, and the following year, Truxtun Beale presented Bakersfield with more architecture. The Beale Memorial Clock Tower, gifted in memory of his mother, Mary Edwards Beale, was built in the middle of the intersection of Seventeenth Street and Chester Avenue—a fact not always fully appreciated by city leaders. Beale had been inspired by a clock tower in Spain. (Courtesy of Kern County Museum.)

17

In 1907, Truxtun Beale donated five acres off of Oleander Avenue and Dracena Street for the construction of Bakersfield's first city park, Beale Park. The donation included materials for landscaping, a swimming pool, and a Greek amphitheater. Truxtun Beale died in 1936. Of his three major donations to the city, only Beale Park remains in its original location. (Courtesy of Kern County Museum.)

Among the city's early leaders was Colonel Baker's business partner and posthumous son-in-law, Henry Jastro, a cattleman and politician. Born Henry Jastrowitz in Germany in 1848, Jastro immigrated to the United States with his family at age 13. He settled in Bakersfield in his 20s, became Baker's business partner in a brewery, and six months after Baker's death, married Baker's 17-year-old stepdaughter Mary. (Courtesy of Kern County Museum.)

Henry Jastro was a major player in civic life. For 50 years, he worked for the influential Kern County Land Company, which owned 1.4 million acres in four states. He was president of the board of supervisors for a quarter-century and owned the Bakersfield Gas & Electric Company. Jastro, who donated land for the city's second public park, Jastro Park, died in 1925. (Courtesy of Kern County Museum.)

By the time Henry Jastro produced California's first cotton crop near Bakersfield in 1885, the city had invented and reinvented itself culturally, architecturally, and economically more than once. Its central district had everything a burgeoning frontier town could hope to offer—not just stables, blacksmiths, grocers, and saloons, but also doctors, dentists, lawyers, and land offices. (Courtesy of Kern County Museum.)

The city's steady progress halted on July 7, 1889, when a fire started in the kitchen of Mrs. N.E. Kelsey, who had been preparing Sunday dinner. The fire burned for three hours, destroyed 196 buildings, killed one, and left 1,500 homeless. The courthouse was among the few buildings that survived. The transition of Bakersfield from frontier town to 20th-century city began in earnest. (Courtesy of Kern County Museum.)

At the time of the 1889 fire, a second wave of Chinese immigrants had started to settle. The first wave, in the 1870s, arrived with the completion of the Transcontinental Railroad. By 1880, Bakersfield had 250 Chinese residents, mostly Cantonese, living in what was eventually known as Old Chinatown (Twentieth to Twenty-second Streets between L and K Streets). They spoke a Cantonese dialect called sam yup. (Courtesy of Kern County Museum.)

By 1890, a second wave of Chinese immigrants had built New Chinatown (Seventeenth and Eighteenth Streets between Q and R Streets). They spoke sam yup, which was just different enough to be obvious to the Chinese. Many worked for Yen Ming, who farmed hundreds of acres on the spot now occupied by the city's Valley Plaza mall. Many Chinatown buildings remain, along with evidence of tunnels that linked basements—some of them opium dens. (Courtesy of Kern County Museum.)

By the 1890s, another distinct minority group had taken root in the community just a mile or so east of Bakersfield. Basques, hailing from the region of northern Spain and southwestern France delineated by the Pyrenees mountains, were willing to take the job few Americans would consider, sheepherding. Like most Basque communities around the country, Bakersfield's was founded near a railroad depot—the Sumner station. (Courtesy of Kern County Museum.)

For generations, young Basques, speaking no English, arrived at the new Southern Pacific station east of Bakersfield, pictured here, clutching scraps of paper bearing their destination. They walked to Basque boardinghouses nearby where they waited for sheepmen to give them work. One such boardinghouse, opened in 1893 by Faustino Mier Noriega and Fernando Etcheverry, was the Iberia Hotel. In 1906, it became the Noriega Hotel. (Courtesy of Kern County Museum.)

The defining event of the early 1900s involved oil. As early as 1864, local entrepreneurs had mined tar in open pits for asphalt and kerosene, but things changed in 1896 when the Shamrock Gusher blew near the town of McKittrick; increasingly, oil wells began to replace tar mining operations. Then came the 1909 Midway Gusher, which helped turn the tiny village of Taft into a boomtown. (Courtesy of Kern County Museum.)

Two

BETWEEN FIRES

During the final decade of the 19th century and the first two of the 20th, Bakersfield looked the part of an Old West town that had come of age. Its ambitious skyline was reminiscent of Sacramento, Portland, and San Francisco: tall, ornate false fronts with arched windows, peaked spires, wood-plank sidewalks, and wide dirt boulevards. This is Chester Avenue looking north in 1895. (Courtesy of Kern County Museum.)

Pre-1889 Bakersfield exists only in photographs, and there are precious few of them. The city's old town hall, built in the early 1870s on Seventeenth Street between Chester Avenue and Eye Street, not far from Colonel Baker's original field, was not a work of architectural artistry, but the two-story building was sufficient to meet the town's needs. It was destroyed in the fire of 1889. (Courtesy of Kern County Museum.)

The original town hall, at far left in this 1875 oval albumen, did not last and neither did the municipality it served. Three years after the city's incorporation in 1873, Bakersfield's town council voted to disincorporate, partly to get rid of its cantankerous, high-handed marshal and partly because city leaders could not collect enough tax dollars to maintain services. Bakersfield reincorporated in January 1898, nearly 22 years later. (Courtesy of Kern County Museum.)

When Colonel Baker first laid out the boundaries for Bakersfield, the town's only fire protection was a single wagon loaded with barrels of water. After significant fires in 1872 and 1873, the all-volunteer Bakersfield Fire Company was reorganized into the Bakersfield Fire Department, whose divisions included the Eureka Engine Company, the Neptune Hose Company, and the Alert Hook and Ladder Co., pictured here in 1881. They too proved inadequate. (Courtesy of Kern County Museum.)

One of the more obvious answers to Bakersfield's challenges with fires was its collective choice of building materials. The time had come to replace wood with brick. However, brick structures, like this general merchandise store operated by Pablo Galtes, were rare. Pictured in 1885, Galtes's store, the first of its type in the town, opened in 1878; others soon followed out of necessity. (Courtesy of Kern County Museum.)

In 1886, James Curran opened the Sandstone Brick Company, which turned out handmade red clay bricks. The 1889 fire razed the city, and demand for building materials skyrocketed. Sandstone expanded to include bricks, lime, cement, wall plaster, lumber, and oil for Bakersfield's reconstruction. One newspaper proclaimed a week later that entire city blocks of brick "will be constructed as soon as men and means can do it." (Courtesy of Kern County Museum.)

The 1889 fire created opportunity in a broader sense as well. In the damage to its finest and most glamorous edifice, the Southern Hotel—pictured here in 1887—leaders saw a way to redefine the city. They imagined a fully redesigned, ever-expanding metropolis with the reconstructed Southern Hotel as its central feature. The Southern, to use its popular abbreviation, stood at the northwest corner of Chester Avenue and Nineteenth Street. (Courtesy of Kern County Museum.)

In 1889, the fire-damaged Southern Hotel was rebuilt in an ornate Victorian style, with three stories and 84 rooms, each with hot and cold running water and natural gas heating. The Southern was said to have been built for a city three times the size of Bakersfield and on a par with the most glamorous hotels in San Francisco. (Courtesy of Kern County Museum.)

In 1892, voters chose between Pres. Benjamin Harrison and challenger Grover Cleveland. The local ballot also asked whether the city should invest in a new high school. Voters sided with history in both cases. Cleveland tallied 1,266 votes to Harrison's 992, and supporters of a new high school prevailed 1,274 to 286. Kern County High School opened January 12, 1893, with 32 students, pictured here. (Courtesy of Kern County Museum.)

In 1894, its second year of operation, the school held its first graduation ceremony, with three students—May Stark, Adaline Nicholson, and Ella Fay, seen here—accepting diplomas before a packed house at Scribner's Opera House. In 1915, the school became known as Kern County Union High School, and in 1945, it became Bakersfield High School, although evidence suggests it was referred to as Bakersfield almost from the beginning. (Courtesy of Kern County Museum.)

Boxing and horse racing were Bakersfield's most popular spectator sports in the early days, but football and baseball were also familiar. Bakersfield's first organized baseball team, pictured here, played against neighboring towns starting in the 1870s. Kern County High School played its first baseball game on January 12, 1893, and by the middle of the next decade, the Drillers were a San Joaquin Valley powerhouse. (Courtesy of Kern County Museum.)

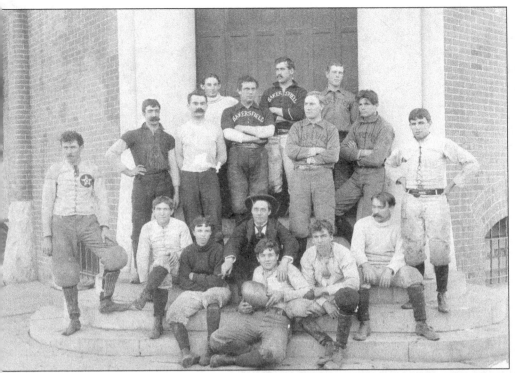

Twenty years after Harvard and Yale played one of the first games of football using codified rules, Bakersfield fielded an amateur team, pictured here, which took on teams in neighboring towns between 1895 and 1902. Many players were undoubtedly alumni of Kern County High School, which played its first football game on February 22, 1893. The Drillers lost that game 90-0 but soon found their footing. The team's seven state championships are the most in California history. (Courtesy of Kern County Museum.)

When Alfred Harrell, then the 34-year-old Kern County superintendent of schools, purchased The *Daily Californian* on January 26, 1897, the newspaper was headquartered in a modest brick building at Twentieth and Eye Streets. The paper started in 1866 as the *Havilah Courier*, but as that town's mineral wealth became depleted, the population—and its newspaper—moved southwest to Bakersfield in 1872. (Courtesy of Kern County Museum.)

Prior to 1899, the land that would become the world-class Kern River Oil Field was a promising but unrealized network of oil seeps. That year, however, substantial pockets of oil were discovered at 70 feet and speculators dug there for oil. By the end of the following decade, the oil field, pictured in 1910, justified its own railroad spur. (Courtesy of Kern County Museum.)

These two well-dressed women outside the Kern County Courthouse in 1903, standing beneath a slightly incongruent 20-foot-tall Canary Island date palm, give some indication of the building's enormity. The courthouse, designed by architect Albert A. Bennett, was built in 1876 but soon proved inadequate for the needs of the growing county, and in 1896, it was reconstructed with additions that doubled its size. (Courtesy of Kern County Museum.)

Bakersfield still had a Wild West element in the early 20th century. From its beginnings, the town attracted every scoundrel imaginable, including claim-stealing ore miners and prostitutes without the means or moxie to venture north to Alaska. Well into the 1880s, sidearms settled mundane grievances long after other similarly sized towns had adopted more refined tastes in entertainment. Jim McKinney, pictured in a deceivingly agreeable moment, was one such scoundrel. (Courtesy of Kern County Museum.)

One of the last Old West shootouts took place in 1903, when fugitive Jim McKinney holed up in this Chinatown joss house. Surrounded by lawmen, McKinney took three slugs to the chest before killing deputy sheriff Will Tibbet and city marshal Jeff Packard. Breathless headlines in Philadelphia and New York made comparisons to more famous gun battles in Virginia City, Abilene, Dodge City, and Tombstone. (Courtesy of Kern County Museum.)

Arsonists burned Railroad Avenue School to the ground shortly after it opened in 1876, and it was not rebuilt until 1901—this time with brick. It was renamed Emerson School in 1904, the year these eighth-grade boys were instructed to open a textbook and look at the camera. Emerson was an elementary school at first, but as the town grew, it became a junior high school. (Courtesy of Kern County Museum.)

"Fold your hands and sit up straight, please." One can almost hear the teacher issue those instructions to Lowell Elementary School's second graders—with girls on the left and boys on the right—on this day in 1904. Lowell School, sometimes called the Tenth Street School, was built in 1902 at the corner of Tenth and H Streets and was torn down in 1954. (Courtesy of Kern County Museum.)

St. Francis Church first offered a storefront Mass to six Catholic families in 1871. In 1910, it opened this stunning, double-spired edifice at Truxtun Avenue and Eye Street, but it was a long time coming. The first building committee included Col. Thomas Baker and Julius Chester, for whom the city's main commercial boulevard, Chester Avenue, is named. Plans were completed in 1881, but nearly 30 years passed before construction began. (Courtesy of Kern County Museum.)

The Standard School District was formed in 1909 after the Standard Oil Company, the forerunner of Chevron Corp., found itself sitting on a veritable gold mine in the nearby Kern River Oil Field. The company donated five acres in what had variously been called over several years, Waits, Oil Center, Oil City, and, ultimately, Oildale. Standard School opened in 1911. (Courtesy of Kern County Museum.)

Bakersfield's Freemasons completed a new lodge on Eighteenth Street south of H Street in 1926, leaving the old, distinctive lodge at Twentieth Street and Chester Avenue to other uses, including, as seen here in 1926, a dentist's office operated by one Painless Parker. The building still stands; however, it lost its two top floors and virtually all architectural flair in the 1952 earthquake. (Courtesy of Kern County Museum.)

Once upon a time, local government officials had a hard time collecting taxes, but life must have been simpler for the Kern County auditor's office when this photograph was taken in 1912, given the success of local oil companies—and the windfall their property tax liability produced for the growing county seat of Bakersfield. (Courtesy of Kern County Museum.)

34

East Bakersfield's cuisine-driven Basque culture lives on today at restaurants like Wool Growers and, until its closure in 2021, the Noriega Hotel, which won a 2011 James Beard Award for culinary excellence. Here, bartenders at the Pyrenees Cafe wait for the evening crowd to pour through its doors in 1939. (Courtesy of Kern County Museum.)

Fire departments around the country entered a new era around 1915 when many of them completed the transition from predominantly horse-drawn fire engines to motorized. That is the approximate date of this photograph showing the Bakersfield Fire Department deploying horse and machine side by side. Fire departments spent more time and money purchasing and caring for one horse than paying 10 firefighters. (Courtesy of Kern County Museum.)

Hochheimer's was the preeminent department store in Bakersfield between 1900 and 1923. The founding Hochheimer brothers, born in Pittsburgh of German descent, opened their first store in the rice-farming town of Willows in 1879. Malcolm Brock, the Hochheimers' nephew, joined the business in 1894, and in 1900, helped them open a store in Bakersfield. It occupied most of a city block on Chester Avenue between Nineteenth and Twentieth Streets. (Courtesy of Kern County Museum.)

In 1919, a huge fire wiped out several businesses, including Hochheimer's. The store relocated and recovered, but a different disaster darkened the company's fortunes in 1922. A rice crop failure caused the Willows store to close. Malcolm Brock bought the troubled family business, and in 1924, rechristened it Brock's—a name many residents remember 30 years after its demise. Brock's was acquired in 1987 by now-defunct Gottschalks. (Courtesy of Kern County Museum.)

Three

INDUSTRY'S FOOTHOLD

In 1924, Ford dealer George Haberfelde—fourth from left, posing with his automotive garage staff—took his daughter on a tour of some of the great capitals of Europe and Asia. When he returned, inspired by the imposing dignity of German, Moorish, and Roman architecture, as well as the works of French-trained American Louis Sullivan, he commissioned architect Charles Biggar to design a building at Seventeenth Street and Chester Avenue. (Courtesy of Kern County Museum.)

The handsome, six-story Haberfelde Building, pictured in 1930, was at the corner of Seventeenth Street and Chester Avenue. Owner George Haberfelde, a German immigrant, was a charter member of the California Automobile Dealers Association and its first president. The one-time Bakersfield mayor was also instrumental in the construction of a one-mile oval racetrack in Bakersfield, which hosted many of the prominent competitive drivers of the day. (Courtesy of Kern County Museum.)

George Haberfelde poses with his sales staff at a 1928 dinner meeting celebrating sales performance. Note the separate sales category posted for Ruckstells, a two-speed rear axle invented in 1913 by Grover Ruckstell, who worked in a west Kern Ford garage. Ruckstell's invention was designed for use on oil field trucks, but by 1928, the axle was an authorized Ford accessory endorsed by Henry Ford himself. (Courtesy of Kern County Museum.)

The Hotel Willis sign is the highest point on the Bakersfield skyline in this photograph of Chester Avenue at Eighteenth Street in 1925. Among the breakthroughs that decade were Pioneer Mercantile selling the city's first air-conditioning unit to Henry Eggers for his clothing store, installation of the first automatic traffic signals in the county on Chester Avenue, and the Hippodrome Theatre screening the city's first "talkie," *The Ghost Talks*. (Courtesy of Kern County Museum.)

Only one building remains from this 1928 photograph looking west on Eighteenth Street toward H Street: the eight-story Spanish Revival–style Padre Hotel, now a popular boutique hotel, bar, and restaurant, completed the same year this photograph was taken. The Padre had an auspicious and flamboyant beginning in the Central Valley's early and notorious oil rush days, but its most compelling stories were yet to be told. (Courtesy of Kern County Museum.)

Construction of Bakersfield's original city hall, pictured in 1926, was championed by businessman Alphonse Weill. In 1909, he declared, "I never appreciated how badly we needed a place for the transaction of the business of the city" until the night he walked to a meeting of city trustees and encountered "the smell of the stable" from the nearby firehouse. Entering the trustees' "inadequate quarters," he said, "convinced me." (Courtesy of Kern County Museum.)

Scribner's Opera House, which opened in 1899, featured more than opera. It hosted boxing matches, graduations, topical lectures, and much more. Nine theaters dotted Chester Avenue in those days, leading some to call the area "Theater Row." In 1920, Scribner's was partially demolished and reconstructed as the California Theatre, shown here. Its final show, in November 1966, was *The Fortune Cookie* with Jack Lemmon. (Courtesy of Kern County Museum.)

The Southern Hotel, pictured on June 10, 1926, stood at the corner of Nineteenth Street and Chester Avenue. The hotel survived the fires of 1889 and 1919 but could not remain economically viable in the postwar years. It was razed and replaced by a JCPenney department store in the late 1940s. (Courtesy of Kern County Museum.)

The Kern County Courthouse was originally the repository of records, but by the mid-1900s, the county needed more room. A Beaux Arts–style hall of records was constructed in 1909 across the street. The building, pictured in 1927, had a design flaw: its rotunda trapped heat. A 1939 remodel changed the style to Public Works Administration Moderne. (Courtesy of Kern County Museum.)

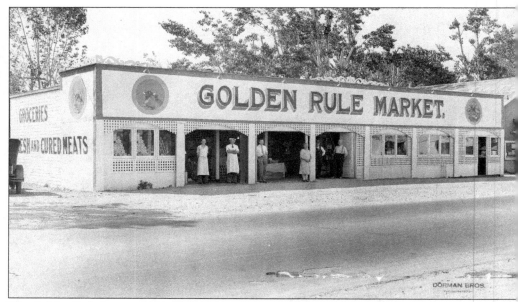

The Golden Rule Store at 2023 Chester Avenue, pictured in 1924, featured fresh and cured meats in its grocery department, but also carried household items. A 1914 newspaper advertisement for the store offers gingham fabric for 7¢ a yard. In case potential customers were not familiar with the location, ads reminded them to be on the lookout for "The Yellow Front!" (Courtesy of Kern County Museum.)

The Golden Rule Store's spacious aisles, pictured in 1924, facilitated an efficient shopping experience. The store's name was common during the first quarter of the 20th century. There was "Golden Rule Sunday" and a "Golden Rule Week." US Secretary of War Newton Baker even cited it in a 1916 foreign policy speech. (Courtesy of Kern County Museum.)

The Bakersfield Fire Department had a four-vehicle fleet in 1922, at the time of this photograph. That was a big change from just a decade before, when the department relied exclusively on horses. The station itself changed over time as well. Its original headquarters burned to the ground in 1904 and was replaced at the same location by this brick building. (Courtesy of Kern County Museum.)

Firefighters hose down the last embers of what had been the JCPenney department store, two doors down from the Grand Hotel, in 1925. In 1939, Bakersfield Fire Station No. 1, designed by architect Charles Biggar with funding from the New Deal's Public Works Administration, opened at 2101 H Street. The station continues to serve the downtown area today. (Courtesy of Kern County Museum.)

Alfred Harrell bought the *Bakersfield Californian* in 1897 and gave the newspaper its present name in 1907. Nineteen years later, he moved the business three blocks south to a Charles Biggar–designed edifice at 1707 Eye Street, which years later was listed in the National Register of Historic Places. Harrell served as editor and publisher until his death in 1946. (Courtesy of Kern County Museum.)

The Kern County Union High School track and field team of 1925 poses following their annual interclass meet, won by the juniors. The school now has a modern, all-weather track, but the grandstand of Griffith Field remains. The school, today known as Bakersfield High School, dominated athletics in California's Central Valley for decades. (Courtesy of Bakersfield High School archive.)

44

Emerson School boys proudly display their awards earned in a 1922 track and field meet. Students in those days only had to attend school 143 days a year. A decade later, the school year went up to 175 days, which is still the standard in the United States. (Courtesy of Kern County Museum.)

Police officers take two unsavory-looking "criminals" into custody as part of a public service demonstration for students in 1925. The challenges of Prohibition and the rising prominence of glamorized mafia figures in the mid-1920s made public relations gestures like this—which depicted criminals as decidedly unromantic characters—seem necessary. (Courtesy of Kern County Museum.)

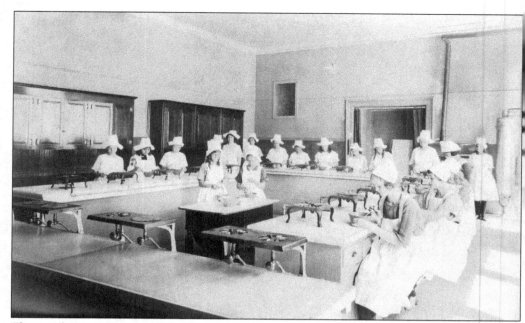

These girls in an Emerson School cooking class in 1922 were required to wear "sanitary" hats reminiscent of nurses' headwear of the time. During the 1920s, by the time they were 25, US citizens had completed a median of 8.2 years of school. Today, 92 percent of the US population has attained a high school diploma by the time they reach 25. (Courtesy of Kern County Museum.)

While the girls were refining their home economics skills, the boys in this 1922 Emerson School manual labor class were learning welding and carpentry. A post–World War I baby boom flooded US schools throughout the 1920s, and industrial growth prompted large businesses to encourage vocational training courses like this one. (Courtesy of Kern County Museum.)

Oil workers monitor the progress of Kern County oil derrick in 1925. US oil production had reached 450 million barrels by this time, prompting unrealized fears that the nation was about to run out of oil. Government officials predicted that the nation's petroleum reserves would last just 10 more years. Until the 1910s, the United States produced 65 percent of the world's oil. (Courtesy of Kern County Museum.)

Major players like Standard Oil were well entrenched in Kern County by 1910, but enough opportunity remained in the vast oil fields for independent speculators to make some farmers wary of intruders. Here, a man and woman pose outside their home, shotguns at the ready, to greet uninvited entrepreneurs, in 1912. Behind them is a collapsed barn and, in the distance, an oil derrick. (Courtesy of Cindy Dodd.)

Oil workers in cowboy gear gather outside the union hall for a group photograph in anticipation of the city's second annual Frontier Days celebration in October 1935. The oil workers had two reasons to celebrate—merchants had decided all workers in the city would be given time off during the two-day celebration, and the oil industry had decided on a production curtailment to boost prices and salaries. (Courtesy of Kern County Museum.)

In the 1920s, Kern County cotton pickers were primarily white, but by the following decade, black farmworkers had started to replace them. Workers dragged long canvas sacks and stuffed them to capacity. Although cotton growers paid marginally better in California than in other states, wages here declined from $1.50 per hundred pounds in 1928 to just 40¢ in 1932. (Courtesy of Kern County Museum.)

A cotton farm manager uses a triple-wide hand truck to move bales of cotton in 1925. Cotton came to Kern County in 1865, when about 120 acres were planted with the crop using Chinese laborers brought from San Francisco. By 1920, a total of 175,000 acres were planted with Upland cotton. (Courtesy of Kern County Museum.)

As early as 1853, California growers planted almonds of European origin. Most were grown in the Sacramento area, but by 1925, almonds had a small foothold in Kern County. Here, workers shell and sort the freshly dropped, fuzzy-shelled crop on a northern Kern County farm. By 1970, much of the state's almond production had moved to the San Joaquin Valley, and today, Kern County is a top producer. (Courtesy of Kern County Museum.)

It was backbreaking labor, but some of these farmworkers, hoes poised for work, managed a smile in 1925. Early Kern County farmworkers included whites, Latinos, Japanese, Chinese, Filipinos, and blacks, and crops included cotton, onions, potatoes, and wheat—but the crops, like the demographics of the workers, have changed over the decades. (Courtesy of Kern County Museum.)

It is not clear how onions came to California, but the first official crop report of 1852 lists a harvest of five million pounds. By 1917, California was an important source of onion seed for the nation, although bulbs were not yet an important export. This Kern County onion field was photographed on May 14, 1926. Onion acreage has remained stable since 1920. (Courtesy of Kern County Museum.)

Four

A City Develops

Tens of thousands of Dust Bowl refugees descended on the West Coast starting in 1935. The impact of those migrants from Oklahoma, Texas, Kansas, and Arkansas—or Okies, as they were collectively and disparagingly known—was felt in Bakersfield more than any other landing spot in California. Here, a family arrives from Oklahoma at the Shafter Labor Camp 20 miles north of Bakersfield. (Courtesy of Kern County Museum.)

The Shafter Labor Camp, a city of 283 tents and small wood-frame buildings on 77 acres, pictured in 1939, was one of the first of the 16 camps built in California. It was the less famous of the two in Kern County after the Weedpatch Camp near Arvin, the setting for John Steinbeck's *The Grapes of Wrath*, but its migrant families dealt with the same challenges. (Courtesy of Kern County Museum.)

In 1939, residents of the Shafter Labor Camp entertain themselves and their fellow campers with a four-piece band. During the second half of the decade, the population of seven California agricultural counties—Yuba, Monterey, Madera, Tulare, Kings, Kern, and San Diego—grew by an average of almost 43 percent, led by Kern with a staggering 63 percent increase. (Courtesy of Kern County Museum.)

The onslaught of migrants so overwhelmed Bakersfield in 1939 that Bakersfield High School was forced to set up overflow classrooms for the children of Dust Bowl refugees. These makeshift classrooms were assigned names such as Tulsa and Oklahoma City, as seen here. Whether the names correlated to the students' original homes or were simply added in jest is lost to history. (Courtesy of Kern County Museum.)

A banner touting "Franklin D. Roosevelt for President" hangs over Chester Avenue in 1932. The election, which took place against the backdrop of the Great Depression, saw Roosevelt, the Democrat, defeat the Republican incumbent Herbert Hoover in a landslide. Roosevelt, the first Democrat in 80 years to win an outright majority, took 58 percent of the vote in California and a whopping 70 percent in Kern County. (Courtesy of Kern County Museum.)

Bakersfield fire chief Phil Pifer, left, and police chief Robert Powers were two of the city's best-known public servants throughout the 1930s and early 1940s. Powers, who got his start as a railroad detective, went on to serve in the postwar administration of Gov. Earl Warren. Powers became an adherent of the Bahá'í faith and later hosted a nationally syndicated radio program, *This I Believe*. (Courtesy of Kern County Museum.)

Mathias Warren, 73, wealthy father of Earl Warren—then the Alameda County district attorney—was murdered in May 1938 as he sat alone in his home one night counting rent receipts, and the crime was never solved. Years later, rumors of a deathbed confession by Ed Regan, who had been Mathias Warren's primary rental property handyman, found its way to investigators, but the case was never conclusively closed. (Courtesy of Kern County Museum.)

Lawrence Tibbett, son of lawman Will Tibbet, killed in the 1903 Joss House shootout, became the New York Metropolitan Opera's first great baritone. Tibbett (who added a "t" to his name) sang leading roles with the Met 600 times from 1923 to 1950. He had a brief career as a screen actor, earning an Academy Award nomination for Best Actor for 1930's *The Rogue Song*, and appeared on the cover of *Time* in 1933. (Courtesy of Kern County Museum.)

The California Highway Patrol (CHP), created by an act of the state legislature on August 14, 1929, set up operations in Bakersfield in 1930, the year of the CHP's first academy graduating class. The agency's Bakersfield office started with these two patrol cars and worked Central California's multiple two-lane state highways. (Courtesy of Kern County Museum.)

The Garcés Traffic Circle was created in 1932 along a rural stretch of US Highway 99, and stood as a stark infrastructural oddity for seven years, until the unveiling of the landmark statue that would give the roundabout its name and its distinction. (Courtesy of Kern County Museum.)

The Father Garcés statue was erected in honor of Spanish missionary and prolific traveler Francisco Garcés (1738–1781). The limestone statue, designed by John Palo-Kangas in 1939, is more than 20 feet tall. It has a near twin in the center of the Long Beach Traffic Circle in Long Beach, California. The pedestal of the Bakersfield statue bears a plaque declaring it a historic landmark. (Courtesy of Kern County Museum.)

The Garcés Traffic Circle is pictured in 1939, looking south down Chester Avenue toward the city's central business district, shortly after the statue's installation. In the coming decades, an overpass was built almost directly over Father Garcés's head, which cluttered the view but made life easier for traffic on Golden State Avenue (Route 204). It stands at the intersection of Chester Avenue, Golden State Avenue, and Thirtieth Street. (Courtesy of Kern County Museum.)

The California Theater celebrated its 18th anniversary in February 1938 with a run of *Gold is Where You Find It*, a Western starring George Brent, Olivia de Havilland, and Claude Rains. The California, built over the partial ruins of Scribner's Opera House, was one of a half-dozen movie houses within a compact, four-block area of the city. (Courtesy of Kern County Museum.)

Sheriff John Caswell Walser's lawmen, pictured here in 1930, made hundreds of arrests, but the one that got away was the most heinous. In December 1927, Walser's deputies tracked down William Edward Hickman, the Los Angeles "maniacal butcher," who murdered a 12-year-old girl and then fled on a red motorcycle. Hickman was cornered in Mojave but somehow escaped, only to be caught two days later in Oregon. (Courtesy of Kern County Museum.)

This Kern County firefighter, rappelling down a building in 1937, worked for one of the finest organizations of its kind in the nation, at least according to a 1938 underwriter's report. The board of underwriters of the *Pacific* praised the department for a number of practices and techniques, including its groundbreaking arrangement with the California State Department of Forestry. (Courtesy of Kern County Museum.)

Kern County firefighters pull hoses from a fire truck in 1937. That year, the fire department had 55 full-time firefighters and expanded to 87 during the summer fire season. The station had 12 firemen, 2 cooks, and 2 stenographers, as well as the offices of Chief Harold Bowhay and his two assistants. (Courtesy of Kern County Museum.)

In 1938, the year of this photograph, the Rio Bravo oil field east of the city was just coming into its own. First tapped by the Union Oil Company, Rio Bravo was the first field in California to reach deeper than 11,000 feet, and for a time, it had the deepest-producing well in the world. By 1974, the field had produced more than 100 million barrels of oil. (Courtesy of Kern County Museum.)

Agriculture and oil, in one form or another, existed side-by-side almost since the day Bakersfield first came into being. Here, a table-grape vineyard, with raisin-drying trays between rows, bears fruit with an oil derrick looming on the horizon east of the city in 1930. (Courtesy of Kern County Museum.)

California's citrus industry started in Southern California, where 85 percent of the state's citrus fruits were produced in the 1920s. By 1932, when this photograph was taken, citrus was making inroads around Bakersfield. After World War II, due to urbanization and the citrus disease tristeza, the crop moved more substantially into the San Joaquin Valley. (Courtesy of Kern County Museum.)

60

Williams School, pictured here in 1932, was one of the Bakersfield City School District's first schools. It served as a community center as well as a grammar school, with a handful of civic organizations holding regular meetings in its cafeteria. Its annual Christmas pageant was among the city's best-loved and best-attended events. (Courtesy of Kern County Museum.)

Shell's Owl Tree service station, pictured in May 1937, was one of Bakersfield's most recognizable fueling stops. Gas was about 20¢ a gallon, or about $4.00 a gallon in 2022 dollars. The national average price did not reach 30¢ a gallon until 1956. (Courtesy of Kern County Museum.)

To meet the growing market for motor fuels, Standard Oil of California came up with the world's first service station in 1907—in Seattle. Before that, motorists bought gasoline from hardware stores, general stores, pharmacies, and even blacksmiths, which already had relationships with refineries, from whom they bought kerosene for resale. By the 1930s, Standard stations were all over the West—including Bakersfield's station No. 200, pictured in 1932. (Courtesy of Kern County Museum.)

The Southern Hotel at Chester Avenue and Nineteenth Street, pictured in 1936, had a little bit of something for everyone—not only the best overnight accommodations in the southern San Joaquin Valley, but also fine dining, a smokers' lounge, and, on Sunday morning, worship services of the Unity Truth Center with Pastor Delia Shutts presiding. (Courtesy of Kern County Museum.)

Chester Avenue showcased the best of what Bakersfield had to offer, architecturally speaking, as this view of the city's main drag looking south from Nineteenth Street in 1938 demonstrates. The primary commercial boulevard was about to add an enduring new structure, the Sill Building, whose plans were made public on May 13 of that year. (Courtesy of Kern County Museum.)

In 1939, the year this photograph of city hall and St. Francis Church was taken, plans were announced for the renovation of the hall of records, seen at center. "A thing of beauty perhaps in 1908 but hardly a joy forever in streamlined 1939," the *Californian* newspaper proclaimed. Architect Frank Wynkoop promised work would commence the month immediately following the announcement. (Courtesy of Kern County Museum.)

Property tax day in November 1938, the year this photograph of Bakersfield City Hall was taken, did not look very Depression-like. Debtors lined up at city hall to beat the deadline and pay their city and county taxes. "When you have so much money you have to pile it in laundry baskets," the *Californian* reported of the haul, "you've got something." Receipts exceeded $4 million, a record. (Courtesy of Kern County Museum.)

Wells Fargo Bank parked one of its trademark stagecoaches in front of the Granada Theater in East Bakersfield just before Christmas 1937 as part of a promotion. Syndicated newspaper gossip columnist and radio news commentator Walter Winchell, starring as himself, was appearing with Simone Simon in *Love and Hisses*. (Courtesy of Kern County Museum.)

Five

STORM CLOUDS

Bakersfield's 440-yard relay team of Harold Matlock, Paul Howard, Jack Farr, and Bob Jackson helped the Drillers dominate the Kern Relays, the biggest annual track and field event in the Central Valley, in March 1942. Matlock was drafted into the Army the following year and served as a platoon sergeant in France, Germany, and Belgium. His nephew Eric Matlock became Bakersfield's first black chief of police in 1999. (Courtesy of Kern County Museum.)

When there was actually something to buy, shoppers headed to Chester Avenue, shown here looking south toward the Beale Clock Tower in 1943. Commerce experienced peaks and valleys as sales fluctuated between droughts caused by wartime shortages and periods of heavy activity caused by shoppers stocking up, and even hoarding, when hard to find items showed up unexpectedly. (Courtesy of Kern County Museum.)

A smartly dressed young woman in a calf-length skirt crosses Chester Avenue at Eighteenth Street, just north of the Beale Clock Tower, photographed as part of a city-sponsored pedestrian safety campaign in 1940. Note the lack of a crosswalk. (Courtesy of Kern County Museum.)

Kern County Union High School students line up to register for classes in August 1941. Within four months, the United States would be at war, and the availability of teachers became a problem almost immediately. By October 1942, fully one-third of the certificated school district employees were in uniform. Some 690 Kern County men and women were killed in World War II. (Courtesy of Kern County Museum.)

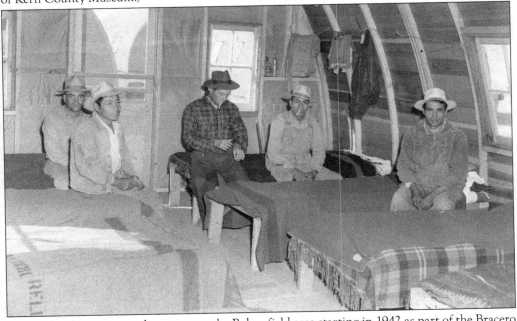

Hundreds of Mexican workers came to the Bakersfield area starting in 1942 as part of the Bracero program, a series of laws and diplomatic agreements that guaranteed them decent living conditions—sanitation, adequate shelter, food, and a minimum wage of 30¢ an hour. Pictured are Braceros in their farm labor housing in Buttonwillow in 1942. (Courtesy of Kern County Museum.)

The line at the Granada Theater is halfway down the block for Cary Grant's *Penny Serenade* in spring 1941. Grant was nominated for the Academy Award for Best Actor for his performance in the film, which told the story of a couple's courtship and marriage with a soundtrack of popular songs relevant to each period of their lives. The Granada, though dilapidated, still stands in East Bakersfield. (Courtesy of Kern County Museum.)

Bakersfield police officers walk the beat in May 1941. There was a lot for them to do. Nineteenth Street, in the vicinity of L Street, especially between the Mint Bar and city councilman Manuel Carnakis's movie theater, had a blocks-long line of second-floor hotels over virtually every ground-floor business—bars, hardware stores and small groceries—and many had brothels. (Courtesy of Kern County Museum.)

Bakersfield police made frequent use of this call box in what they referred to as the tenderloin district—L Street between Nineteenth and Twenty-first Streets—where prostitution and illegal card rooms were rampant. On Saturday nights, they parked an on-call paddy wagon—a 1935 Studebaker panel truck nicknamed "Black Mariah"—next to the call box. (Courtesy of Kern County Museum.)

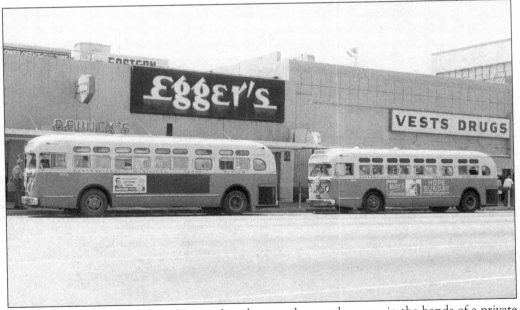

Public transportation in 1941, the year this photograph was taken, was in the hands of a private company called the Bakersfield Transit Company, successor of the Bakersfield and Kern Electric Railway. The company fell on hard times after World War II, and in 1956, the city purchased the bus line. Losses continued, however, and in 1973, voters approved the creation of the independent Golden Empire Transit bus line. (Courtesy of Kern County Museum.)

Here is a street scene from 1941 in front of the Hotel Vernon. Chester Avenue was the city's primary shopping district, with department stores, specialty shops, theaters, and five-and-dimes aplenty throughout the 1940s. The only real competition came from Baker Street, about a mile east. (Courtesy of Kern County Museum.)

The California League began play in 1941 at brand-new Sam Lynn Ballpark. Bakersfield was a charter member, fielding this group—the Bakersfield Badgers. The Badgers finished third that season but played just 67 games the following year before play was suspended because of the war. When the California League resumed in 1946, the Bakersfield team was called the Indians because of its affiliation with Cleveland's major-league team. (Courtesy of Kern County Museum.)

Driller fans pack the stands at Griffith Stadium on November 11, 1941, for the annual Armistice Day Football Classic against Loyola of Los Angeles. Loyola won 14-7, avenging its 41-0 spanking at the hands of the Bakersfield boys the previous year. The loss was Bakersfield's first of the season and only its second in three years. (Courtesy of Kern County Museum.)

Loyola, using the "T" formation with great effectiveness, scores the go-ahead touchdown as Bakersfield players give chase in vain during the annual Armistice Day Football Classic. The non-league game, which at one time pitted Bakersfield against Long Beach High School, dates to at least the 1930s. Virtually every game in the series was played before a loud, packed house at Griffith Stadium. (Courtesy of Kern County Museum.)

The southwest corner of the Hopkins Building at Nineteenth Street and Chester Avenue featured See's Candies, Al's Coffee Shop, Strauss Jewelers, and Kirby's Shoes in this photograph from December 1941. The upper floors were offices for mostly attorneys but also a dentist's office, a public relations firm, and a jeweler, among others. (Courtesy of Kern County Museum.)

California governor Earl Warren, who grew up in Bakersfield, accepts a sack of commemorative potatoes from the Kern County Farm Bureau in honor of the county surpassing $1 billion of potato crop value in 1943. The previous year, then attorney general Warren, a Republican, had defeated incumbent governor Culbert Olson, taking just under 57 percent of the vote. Warren's victory immediately made him a national figure. (Courtesy of Kern County Museum.)

Children line up at Franklin School to purchase Defense Stamps, paying with class credits earned for schoolwork, on January 9, 1942, just 33 days after Pearl Harbor. Classrooms expressed their patriotism and became involved in the war effort by selling and buying Defense Stamps that could be used to purchase US Defense Savings Bonds. (Courtesy of Kern County Museum.)

Students work in the Standard School victory garden, which received a county award for efficiency on March 9, 1944. The government encouraged people to plant victory gardens not only to supplement their rations but also to boost morale. They were used along with ration stamps and cards to reduce pressure on the food supply. (Courtesy of Kern County Museum.)

Local military and city and county officials observe a downtown parade from a city hall balcony on Army Day, April 6, 1942. Various units of military and civilian defense forces participated in the parade, which the *Californian* described as a "grim" display. That day in the Philippines, Japanese forces closed in on Bataan. (Courtesy of Kern County Museum.)

East Bakersfield High School, which opened in 1938, giving Bakersfield its second public high school and Kern County its third, after Shafter High (1928), is shown in 1943. The Blades were the Bakersfield Drillers' chief rival through the 1960s, when the Kern High School District went from three schools to ten. (Courtesy of Kern County Museum.)

In 1943, the year of this photograph of an East High woodworking class, several 18-year-old students left school to fight in World War II. At least two of them were killed and have campus sites named in their honor: Permenter Field, named for Ray Permenter, who died at age 19, and Bayless Hall, named for former student body president Herbert Bayless. (Courtesy of Kern County Museum.)

Apron-wearing girls prepare food in their East High home economics class in this 1943 photograph. As the second-oldest high school in Bakersfield, it opened in September 1938 to freshmen and sophomores only and had a 22-member faculty. The auditorium and gym were not finished until 1940, and there was no cafeteria until 1949. (Courtesy of Kern County Museum.)

Gov. Earl Warren returns to his high school alma mater in Bakersfield for a speech in 1948, six years after he became the driving force behind the internment of more than 100,000 Japanese Americans without any charges or due process. In his later years, Warren expressed regret for his role in the internment, which lasted until the war's end. (Courtesy of Kern County Museum.)

Movie-goers line up outside the Fox Theater in May 1944 for *The White Cliffs of Dover*, starring Irene Dunne as an American-born World War II nurse in a British hospital who is awaiting the arrival of 300 wounded men. Five weeks after the film's release, Allies would embark from Dover and other ports in southern England to storm the beaches of Normandy. (Courtesy of Kern County Museum.)

Six

TRIUMPH

The war ended in the late summer of 1945 when Japan accepted the Allies' terms of surrender. Here, two boys walk down ticker-tape-strewn Chester Avenue on August 14, 1945, peddling copies of an extra edition of the *Bakersfield Californian*. Victory over Japan Day would officially be celebrated on the day formal surrender documents were signed aboard the USS *Missouri* two weeks later. (Courtesy of Bernie Dodd Collection/Cindy Dodd.)

Cars wait for a Santa Fe Railroad train to clear the tracks and make way for traffic on Chester Avenue in November 1940. Bakersfield's population that year was 29,252. Only six California cities had populations of 100,000 or more: Long Beach, Los Angeles, Oakland, Sacramento, San Diego, and San Francisco. Civilian war industries and the Dust Bowl migration would soon change that equation for Bakersfield dramatically. (Courtesy of Kern County Museum.)

The Victorian bandstand at the Oleander-Sunset district's Beale Park has hosted open-air summer band concerts for decades, and does so even to this day, but all was quiet when a driver—perhaps a young friend of photographer Bernie Dodd—parked his Ford convertible on the lawn at the venerable public park in 1945. (Courtesy of Bernie Dodd Collection/Cindy Dodd.)

Exhausted Bakersfield celebrants wait for a city bus in front of the Sill Building at Eighteenth Street and Chester Avenue after a September 1945 World War II victory parade, whose celebratory leaflets and bunting litter the street in front of the busy downtown drug store. Kern County, like communities across the United States, gave all it had to the war effort, not only in terms of fighting men and women—10,881 of them, 690 of whom were killed—but also output from its farms, dairies, livestock ranches, and oil fields. Kern County hosted pilot training at Minter Field, housed 3,000 German (and a few Japanese) prisoners of war in five camps, manufactured parts for defense industry projects, and even developed a rubber mill to help counter shortages created when Japan began to occupy the primary rubber producing countries of southeast Asia. The Bakersfield manufacturing plant derived rubber from parthenium argentatum, or Guayule, a shrub native to northern Mexico and southern Texas. (Courtesy of Bernie Dodd Collection/Cindy Dodd.)

Two sailors celebrate their homecoming in September 1945. Eight million men and women from every service branch, scattered across 55 theaters of war spanning four continents, came home over a period of 360 days, from September 6, 1945, until September 1, 1946. The effort was called Operation Magic Carpet, and was the largest combined air and sealift ever organized. (Courtesy of Bernie Dodd Collection/Cindy Dodd.)

Proclaiming "Victory of the Democracies," the American Legion's 1945 V-E and V-J Armistice Day Parade attracted more than 90 entries, including these soldiers rolling down Chester Avenue in a convertible. More than 20,000 people—nearly a third of the city's population—came out to watch the procession, which was more than a mile long. (Courtesy of Kern County Museum.)

GIs rest in a railcar on their way home to Bakersfield in 1945. Operation Magic Carpet, the government's postwar repatriation program for its overseas fighting forces, transported 22,222 Americans home every day for nearly one year straight—eight million men and women in all. The war department had GIs all over the world—50 percent in Europe and 33 percent in the Pacific, in addition to another 17 percent who were already home at war's end. They came home in Liberty ships, Victory ships, and troop transports—more than 370 craft in total, including aircraft carriers, battleships, hospital ships, and large numbers of assault transports. The undertaking had been in the planning stages since at least 1943, and a prime consideration, in addition to the mobilization of ships, was the development of adequate ports, docking facilities, and demobilization camps after the soldiers, sailors, marines, and airmen reached America's shores. Trains and buses represented the final leg of the journey home for many. (Courtesy of Bernie Dodd Collection/Cindy Dodd.)

East Bakersfield High School seniors enjoy the 1945 prom. Members of that class included Don Rodewald, who for 17 years hosted *The Afternoon Show* on local television. Rodewald, who served as 1944–1945 student body president, enlisted in the Navy upon graduation and served until the end of the war. After retiring from television, Rodewald joined the faculty at Bakersfield College. (Courtesy of Bernie Dodd Collection/Cindy Dodd.)

As big as the 1945 Armistice Day Parade was, the Victory over Japan one-year anniversary parade of September 2, 1946, was bigger, with 30,000 residents watching the solemn, two-mile-long procession. Martial music was provided by the Sixth Army headquarters band, and the crowd witnessed a dramatic reenactment of the raising of the US flag on Iwo Jima. (Courtesy of Kern County Museum.)

A jubilant crowd wraps around the north side of the Beale Memorial Clock Tower to watch the Victory over Japan one-year anniversary parade on September 2, 1946. "A year to the date after Japan capitulated to American might and arms," the *Californian* reported, "and in the midst of general world unrest, Kern county residents today lined the streets of Bakersfield's downtown section and watched a 2-mile parade of marching units, beautifully decorated cars and rhythmically performing musical groups, honoring the first anniversary of the end of organized hostilities, the second V-J Day. . . . More than 200 participants completed the list of participants, sponsored by practically every city and county fraternal, government and private organization. It is estimated that 30,000 persons lined the route of the parade to see the spectacle." City officials were forced to reroute all downtown parades after Chester Avenue was redesigned with a tree-lined center median in the 1990s, but for decades, until after the 1952 earthquake, this curious queue around the clock tower was the norm. (Courtesy of Kern County Museum.)

Uniformed young women, assembled just south of the Bakersfield Californian Building, show off their float featuring the American Legion flag and a floral Liberty Bell and prepare to march at the 1945 Armistice Day Parade, which celebrated American sacrifices in two world wars. (Courtesy of Kern County Museum.)

Among the thousands of Americans who moved from the Midwest and Southwest during the Dust Bowl and World War II eras were citizens of Missouri. When they reached Bakersfield, they formed—starting in the mid-1930s—the Missouri Club, a social and cultural organization that thrived for decades. Here, in 1949, female members gather for a baked goods fundraiser. (Courtesy of Bernie Dodd Collection/Cindy Dodd.)

84

Seven

HEALING AND PROSPERITY

The postwar years brought unparalleled growth to the nation and to Bakersfield. Union Avenue, which already had the pre-freeway designation of US 99, came to life with hotels, nightclubs, and restaurants. In 1945, Oscar and C.L. Tomberlin opened the Bakersfield Inn, and in August 1949, they added its most distinctive feature, the Bakersfield arch, a 200-foot-long footbridge that spanned Union Avenue. (Courtesy of Kern County Museum.)

"Hey look, fellas," one can almost hear this young man say to his friends. "This guy's taking our picture!" This scene from a local barroom captures the workaday fashions and demographics of Bakersfield in 1947, at least for men. Among the patrons, a young soldier and an oil field worker order up. Above them are an advertisement for Acme Beer and two pinup girl calendars. (Courtesy of Bernie Dodd Collection/Cindy Dodd.)

No one was taking Christmas for granted in 1947, with memories of difficult Christmases past still fresh. Here, Chester Avenue is decorated for the annual Christmas Parade. The best float award went to Summer Circle No. 107, Order of the Druids. Within a month, anti-American unrest was flaring up in Shanghai, China, and Hanoi, Vietnam. (Courtesy of Kern County Museum.)

The Plunge opened in 1916 as the largest outdoor concrete swimming pool in the West—a 100-by-281-foot pool situated less than 100 yards from newly opened US 99. That first year, 1,000 elbow-to-elbow swimmers visited daily. By 1946, the year of this photograph, the Plunge was a regular summer afternoon activity for teens. Five generations enjoyed the Plunge before it closed in 1993. (Courtesy of Bernie Dodd Collection/Cindy Dodd.)

This radio deejay, known to the public only as the Night Watchman, took telephone requests and played popular music five nights a week from 10:00 p.m. to 1:00 a.m. from 1949, the year of this photograph, until late 1953, when television came to Bakersfield. (Courtesy of Bernie Dodd Collection/Cindy Dodd.)

Bakersfield will soon have the finest fire alarm system in the West, Bakersfield fire chief Phil Pifer announced in July 1947, about the time this photograph was taken of a city firefighter shooting water from a hydrant. By September, Pifer said, Bakersfield would have 67 alarm boxes per square mile, better even than San Francisco, which had 56 per square mile. (Courtesy of Kern County Museum.)

Defense attorney Wiley Dorris, right, planned to run for the state assembly in 1918 but was drafted and sent to fight in World War I. His wife, Grace Storey Dorris, ran in his place and was elected, becoming one of the first women in the United States elected to public office—five years before women were permitted to vote. Her husband remained a colorful local character throughout the 1950s. (Courtesy of Kern County Museum.)

All is quiet on a summer night in downtown Bakersfield in 1950, looking down from a Chester Avenue rooftop on the Sill Building, the Southern Hotel, and the California Theater. By 1950, the population of Bakersfield had reached 112,000. California's population had reached 10 million. (Courtesy of Kern County Museum.)

If it was a Saturday night, it was drive-in burger night for many Bakersfield teens, including these two clowning around inside the Budge Inn in 1950. Other popular pre-McDonalds, pre-Burger King burger joints included Michener's, Andre's, Stan's and Ken-Ken's—places for friends to meet up and couples to court. The Budge Inn on Chester Avenue still stands; it is now a mobile phone store. (Courtesy of Bernie Dodd Collection/Cindy Dodd.)

Harvey Auditorium, designed by Charles Biggar for the campus of Bakersfield High School, came to be used often by the general public. Work on the three-story auditorium began in 1940 but was suspended in 1942 due to the wartime rationing of building materials. It was completed following the war, in October 1948—two years after Biggar's death in 1946. (Courtesy of Kern County Museum.)

Supervisors with DiGiorgio Farms, a major Kern County grower throughout much of the 20th century, gather for lunch in 1959. Founder Joseph A. DiGiorgio is at far right, with superintendent Joe Little to his right. DiGiorgio, born in Sicily in 1874, built DiGiorgio Fruit Corporation into what was for a time the largest fruit grower in the world. (Courtesy of Kern County Museum.)

Four boys in an East Bakersfield High School surveyor's class take turns peering through the viewfinder of a total station camera in 1950. The postwar building boom created great demand in the construction trades, so these students were making wise choices in their academic pursuits. (Courtesy of Kern County Museum.)

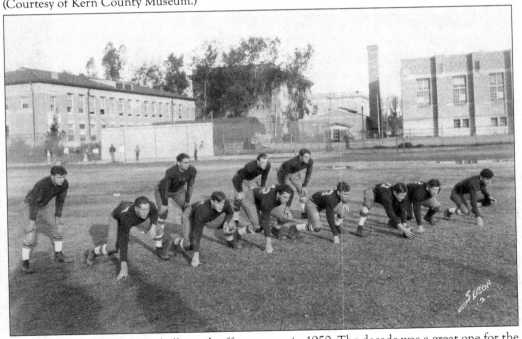

The Bakersfield College football team's offense poses in 1950. The decade was a great one for the Renegades, who won national community college championships in 1953 and 1959 and Potato Bowl titles in 1952, 1955, 1957, and 1958. (Courtesy of Kern County Museum.)

The Kern County Courthouse was built in 1912 at the corner of Truxtun and Chester Avenues, facing west toward Chester. Pictured here in 1946, the courthouse had the distinguished presence of the Greek Parthenon. It was not just a courthouse—the building housed most of county government, including, in the basement, a second incarnation of the library donated by Truxtun Beale. (Courtesy of Kern County Museum.)

Kern General Hospital, forerunner of Kern Medical, was another design project by prolific architect Charles Biggar. The hospital, shown here in 1946, has been substantially expanded and modernized and is still in use today as the county's primary medical facility. (Courtesy of Kern County Museum.)

Eight

THE GREAT EARTHQUAKE

At 4:46 a.m. on July 21, 1952, Kern County shook for 45 terrifying seconds, leveling roofs, crumbling facades, and sending 100-foot-tall water tanks tumbling to the asphalt like slain Goliaths. The third-most powerful earthquake in recorded California history, then and still, was particularly devastating to tiny Tehachapi, 45 minutes southeast of Bakersfield, whose modest downtown was decimated, as this crushed car illustrates. (Courtesy of Kern County Museum.)

A police officer walks past Tehachapi's Juanita Hotel, which lost second-story walls on three sides, leaving guests' beds hanging precipitously over the edge the morning of July 21, 1952. No serious injuries were reported at the Juanita despite the harrowing close call, but the bizarre nature of the damage assured the hotel's eventual loss to the wrecking ball. People felt the earthquake all the way from Winnemucca, Nevada, 600 miles north, to Baja California, 300 miles south. Tall buildings swayed in Phoenix, Arizona. The epicenter was determined to be in a Kern County cottonfield along the White Wolf fault. The magnitude 7.3 earthquake, centered near Tehachapi, killed 11 people in that mountain village and a 12th in the equally devastated farming town of Arvin. The 1952 Kern County Earthquake was actually a series of violent seismic events spanning 33 days, and as bad as the initial shock might have been, the worst, in terms of structural, economic, and cultural loss, was yet to come. (Courtesy of Kern County Museum.)

Thirty-three days after the Tehachapi earthquake, on August 22, 1952, at 3:41 p.m., the most violent and jolting of the aftershocks hit, this one centered in Bakersfield. Edna Belle Ledbetter, 26, was shopping with her younger sister Lily Hobbs at Lerner's Dress Shop on Nineteenth Street, one door east of Chester Avenue, when the building started shaking violently. Lily made it out alive. Edna did not. (Courtesy of Kern County Museum.)

When the magnitude-7.5 Bakersfield aftershock hit, Lerner's Dress Shop, customers screamed; teen girls ducked under dress racks. Edna Ledbetter made a dash for the door and had almost reached it when a side wall came down, crushing her beneath tons of bricks and debris. Her sister was injured but survived. In the eerie aftermath, the only things standing were store mannequins. (Courtesy of Kern County Museum.)

A mile east, the roof of the Kern County Equipment Company came crashing down. Patmar Cozby, 66, a former Southern Pacific Railroad engineer, fell through the collapsing floor into the basement and was crushed under a mountain of falling bricks and farm equipment. He and Edna Letbetter were the only fatalities of the Bakersfield aftershock, but hundreds were injured, and dozens were hospitalized. (Courtesy of Kern County Museum.)

The signature insult of the August 22 earthquake, for many, was the fallen spire of St. Francis of Assisi Catholic Church, one of the most beautiful and distinctive structures in downtown Bakersfield. The sight of one of its twin steeples lying crushed in the middle of Truxtun Avenue drove home the random cruelty of nature in a way no other architectural loss could have. (Courtesy of Kern County Museum.)

Crews worked for weeks to repair the Tehachapi Grade train tunnels; here they are on July 30. One tunnel was discovered to be eight feet shorter than it had been before the July 21 quake. Making the repair work more tedious was the fact that workers dropped their tools and scrambled out into the daylight every time an aftershock rattled the earth—and there were many. (Courtesy of Kern County Museum.)

The Beale Memorial Clock Tower, which since 1904 had occupied the intersection of Seventeenth Street and Chester Avenue, was condemned. The city council had voted to remove it in 1912 anyway to make way for a streetcar system, but backtracked after public outcry. Now, the city council had its justification. Even then, preservationists claimed the damage to the clock tower was cosmetic; they lost the argument. (Courtesy of Kern County Museum.)

97

Devastation was evident all over Bakersfield and beyond, but especially in the city's downtown core, where fanciful architectural embellishments and large windows were everywhere. Three doors down from Lerner's Dress Shop, the Kress department store was in shambles inside and out, with broken glass and fallen masonry strewn along Nineteenth Street. (Courtesy of Kern County Museum.)

Damage to the Hopkins building, where some of the city's best-known attorneys had offices, was severe but ultimately not bad enough to justify the wrecking ball. However, the top three floors of the four-story office building, built in 1904 and rebuilt after the 1919 fire, were deemed unsafe for occupancy, leaving three ghost floors with the names of their tenants still spelled out on decaled glass doors. (Courtesy of Kern County Museum.)

Military reservists were called to duty and police officers were assigned 12-hour shifts, with all vacations canceled. For three weeks, the entire downtown area was blocked off to traffic, both vehicular and pedestrian. Bakersfield counted its blessings, realizing, from the number of heavy masonry bricks like these lodged in newly cracked sidewalks, it could have been worse. (Courtesy of Kern County Museum.)

Several government buildings were deemed lost. One was Bakersfield City Hall, shown here undergoing demolition. Workers had to use cutting torches to bring it down after finding heavy steel reinforcements within. City Council, which had been meeting in the courthouse since the July quake, moved in August to a new meeting place at Carpenters Hall on Twentieth Street. (Courtesy of Kern County Museum.)

The *Californian* building suffered extensive damage, perhaps payback for columnist Jim Day's earlier, well-intentioned levity. He wrote about a couple who, awakened by the predawn July 21 quake, burst outside naked and only after sprinting some distance stopped and realized what they had revealed to the world. Another couple, Day wrote, was spotted sitting on their still-vibrating front lawn with a birdcage—and its resident bird—situated protectively between them. (Courtesy of Kern County Museum.)

The day after the August aftershock, a group of bearded, robed volunteers, here chatting casually with Red Cross workers, showed up to help. They and their guru, a man named Krishna Venta (center) drove up from the Santa Susana Pass of Ventura County. Six years later, two disillusioned cult members assassinated Venta with dynamite, and a decade after that, surviving members hosted an inquisitive Santa Susana neighbor—Charles Manson. (Courtesy of Kern County Museum.)

100

The demise of the Kern County Courthouse, built in 1912, was tragic, but it was not a total loss. The courthouse's four striking limestone Warrior Maidens—each eight feet tall and positioned 65 feet above the street—were saved. Workers carefully removed the maidens with cranes, and they were preserved for some later purpose, which turned out to be their own exhibit at the Kern County Museum. (Courtesy of Kern County Museum.)

Retailers and government offices alike set up shop in tents after their buildings were compromised by the earthquake. Montgomery Ward's moved into tents on F Street, and Brock's Department Store did the same. City and county governments, suddenly needed more than ever, set up tent headquarters at the fairgrounds. Superior Court judges even held criminal trials in public spaces, including city parks. (Courtesy of Kern County Museum.)

Not even Kern County's Beale Memorial Library was going to be defeated by the earthquake. Librarians turned to bookmobiles and tents to supply readers with literature in what was still, for most families, the pre-television days. The library's main branch had been in the basement of the destroyed courthouse. A few years later, the Beale would move into what is now the Jury Services Building. (Courtesy of Kern County Museum.)

Bakersfield mayor Frank Sullivan, left, had to make some hard decisions in consultation with state and county officials. They opted to tear down historically significant, classically designed buildings that might have been salvageable, and replace them with the more mundane rectangles typical of 1950s and 1960s architecture. Then, they plastered over surviving buildings to hide their original character to better match the new structures. (Courtesy of Kern County Museum.)

Passersby on the Chester Avenue side of the Hopkins Building miraculously avoided injury when thousands of pounds of brick and mortar crashed through marquees and awnings onto the sidewalk. It was the second catastrophe for the Hopkins on that date—August 22. Thirty-three years earlier, to the day, the Hochheimer fire of August 22, 1919, destroyed the popular department store, which occupied the Hopkins Building. (Courtesy of Kern County Museum.)

Buildings and sidewalks suffered the most obvious and grievous damage, but dozens of cars were also dented by falling debris. No serious automobile accidents were thought to have been caused by the jolt, although many drivers reported a similar sensation as they traveled down the street: they first thought they had suffered one or more flat tires as the ground shook. (Courtesy of Kern County Museum.)

In 1959, a new Kern County Courthouse opened where the previous one had been, except it faced onto Truxtun Avenue instead of Chester Avenue. Their architectural styles could not have been more different. The ornate 1912 courthouse harkened back to Greek architecture, while the 1959 version was designed in the sharp, rectangular style of the postwar era. It became a model for others to follow. (Courtesy of Kern County Museum.)

Bakersfield's post-earthquake city hall, built in 1954, is still in operation—although many services, including the city manager's office, moved into City Hall North, across Truxtun Avenue, in the early 2000s. The new city hall building, a former law office built in 1995, is ironically reminiscent of the city's pre-earthquake dignity. (Courtesy of Kern County Museum.)

Nine

BAKERSFIELD NOIR

In August 1949, Bakersfield police chief Horace Grayson, center, was accused of taking bribes from bookmakers and madams. The civil service board ultimately cleared him, but two months later, city manager Carl Thornton tried again, accusing the belligerent, autocratic chief of incompetency and dishonesty. The testimony was highly entertaining, but in the end, the civil service board cleared Grayson a second time. Vice flourished anew. (Courtesy of Kern County Museum.)

The Bakersfield Police Department rolled out a new fleet of Indian patrol motorcycles in 1942, including a version with the now rarely seen sidecar, and these officers were more than happy to model them. Indian motorcycles, in their original incarnation, were manufactured until 1953. This new fleet was one of the first major acquisitions by Chief Horace Grayson. (Courtesy of Kern County Museum.)

Bakersfield police officer C.R. "Butch" Milligan, third from left, is front and center in this 1946 photograph of the agency's motorcycle division. Milligan, who would later lead the department's vice unit, holstered a pearl-handled revolver, favored Cadillacs, and boasted of friendships with the likes of actor John Wayne. He once caused a traffic accident by simply walking down Chester Avenue with Clark Gable. (Courtesy of Kern County Museum.)

Chief Grayson, right, presents Lt. Butch Milligan with a commendation for apprehending Jimmy Lee Smith, one of the two suspects in the kidnap-murder of Los Angeles Police Department officer Ian Campbell, a 1963 crime documented in the Joseph Wambaugh bestseller *The Onion Field*. Accomplice Gregory Powell fled back to Los Angeles and was quickly nabbed, but Smith tried to hide out in the Lakeview area of Bakersfield. (Courtesy of Kern County Museum.)

Legendary attorney Morris Chain, founder of a law firm that, through a half-dozen name iterations, still thrives today, poses with former heavyweight boxing champion Joe Louis at a 1952 fundraiser in Bakersfield for John Sparkman, Democratic presidential candidate Adlai Stevenson's running mate. Chain and his chief investigator, Leonard Winters, had uniquely free access to the police department's criminal data files. (Courtesy of Kern County Museum.)

In what some have called the O.J Simpson trial of its day, Morris Chain headed the defense team for Donnell "Spade" Cooley, a Western swing bandleader and one of the first television stars of postwar Los Angeles. In April 1961, Cooley savagely and fatally beat his wife, Ella Mae Cooley, at their Mojave Desert ranch. Daily headlines blared the tawdry details of Cooley's Bakersfield murder trial nationwide. He was convicted. (Courtesy of Kern County Museum.)

Milton "Spartacus" Miller moved to Bakersfield from Chicago in 1954 at age 40 to "oversee his investment" in the Padre Hotel, a brooding, beige-gray, 1928-vintage, gargoyle-appointed edifice at H and Eighteenth Streets in downtown Bakersfield. As two-thirds owner and trustee for the monolithic, eight-story hotel's Chicago-based ownership group, he stayed for the rest of his life. Today, the renovated Padre is a popular boutique hotel. (Courtesy of Kern County Museum.)

This 1961 photograph of the Padre Hotel reflects the angst of the era. The Soviet Union had demanded the withdrawal of all US-allied forces from Berlin and deployed its military across the divided city. The Soviets then erected that enduring symbol of Cold War paranoia, the Berlin Wall. Americans started building fallout shelters, and Milton "Spartacus" Miller made his thick-walled hotel available to all in case of nuclear war. (Courtesy of Kern County Museum.)

Padre Hotel owner Milton "Spartacus" Miller points derisively at his downtown neighbors, whom he believed had allowed the city's central district to decay. He saved most of his wrath for city council, however, by positioning a mock "US Army" missile on the roof and—according to the apocryphal story, which he thoroughly enjoyed hearing but privately denied—pointing it at city hall, some 500 yards away. (Courtesy of Kern County Museum.)

Charles Dodge applied for Bakersfield police chief in 1946 and was passed over. When he tried and failed again in 1966, he ran for sheriff and defeated the incumbent in a landslide. Dodge, campaigning for reelection four years later in this photograph, won again, then retired, and married Lt. Mary Holman, the city's first female police officer. The Dodges both lived into their late 90s. (Courtesy of Kern County Museum.)

Bakersfield High School football star Frank Gifford went on to play one season at Bakersfield College and three at the University of Southern California, where he was named an All-American. He was a sensation in his rookie season with the New York Giants in 1952, scoring seven touchdowns, and was the league's Most Valuable Player in 1956, as the Giants won the National Football League championship. (Courtesy of Kern County Museum.)

Frank Gifford moved easily from a Hall of Fame career in the NFL to that of sportscaster, eventually landing on ABC's *Monday Night Football*, where he was a fixture alongside Howard Cosell from 1971 to 1988. Here, he interviews Mickey Mantle of the New York Yankees in 1965. (Courtesy of Kern County Museum.)

The Bakersfield Boosters, an affiliate of the Philadelphia Phillies, played one season in the California League in 1956. Here, a mounted, uniformed mascot makes an appeal for fans to attend that night's game at Sam Lynn Ballpark. The team finished an abysmal 48-92, but two players, Red Lynn and Jim Golden, went on to play in the majors. (Courtesy of Kern County Museum.)

From the late 1940s to the mid-1950s, the US Atomic Energy Commission, a forerunner of the Department of Energy, encouraged people to search for and mine uranium in rural America. Here, get-rich speculators gather at the Haberfelde Building, where commission officials were preparing to pass out maps of lands thought to hold uranium reserves. (Courtesy of Kern County Museum.)

The uniforms looked authentic enough in this Junior Baseball Association game in 1954. Founded in 1947, the league moved to a patch of land behind Sam Lynn Ballpark in 1951. The diamonds bustled with kids—some 1,600 per season—including future major leaguers Johnny Callison, Steve Ontiveros, George Culver, and Rick Sawyer. The association merged with Little League in the 1990s after its numbers dropped below 800. (Courtesy of Kern County Museum.)

A cab driver picks up a fare at the Bakersfield Hotel, at the corner of Nineteenth and M Streets, in 1955. After the city's streetcar service ceased operation in 1942 and before the Golden Empire Transit bus system was created, taxis dominated the city's decaying and inefficient city-owned bus line in the competition for riders. Not that the Bakersfield Transit Company bus line had not tried. In 1950, the company said it was prepared to make strides as the preferred mode of public transportation for "a growing city," estimating it would carry 4.5 million riders that year and travel 114,000 miles per month. Buses first came to Bakersfield in 1915, offering competition to electric streetcars and bus-taxi hybrids known as jitneys. During World War II, the city's Nineteenth Street track was sold to a reclamation company, which re-sold it for scrap to US war industries, effectively ending the era of streetcars. Some of the city's surviving streetcars went to Nova Scotia, and others turned up as hot dog stands or farm labor housing. (Courtesy of Kern County Museum.)

For teens like these Bakersfield High School students in 1956, out-of-school hours were consumed by breakout singing sensation Elvis Presley, who had a trifecta of hits in "Don't be Cruel," "Blue Suede Shoes," and "Hound Dog." Actress Grace Kelly perpetuated every girl's princess fantasy that year by marrying Monaco's Prince Rainier III, and one in three high school graduates were suddenly going to college. (Courtesy of Kern County Museum.)

In 1965, the Beale Clock Tower was gone, razed in a post-earthquake zeal for urban renovation, but the Haberfelde Building still had a familiar neighbor. The gas station owned by Standard Oil of California, precursor of Chevron, was still across the street to the south. Even the gas station was gone within a few years, removed to make way for the city's Bank of America tower. (Courtesy of Kern County Museum.)

Ten

SUN, FUN, STAY, PLAY

Since 1939, through times difficult and triumphant, the gracefully benign figure of Fr. Francisco Garcés has stood watch at the roundabout intersection of Chester Avenue and Golden State Highway, shown here in time-lapse, looking south, in 1960. At the outset of that decade, Bakersfield had a population of 143,000, a limited geographic footprint, and a bustling downtown that went quiet on rainy evenings. (Courtesy of Kern County Museum.)

Country music came into its own as a commercially viable genre in the 1960s, and two Bakersfield artists led the way. The No. 2-selling country artist in the United States that decade was Merle Haggard (pictured), a homegrown talent, who saw 9 of his first 10 albums land in the Top 10. No. 1 was Buck Owens, a Bakersfield transplant, who notched 26 Top 5 singles, including 16 straight No. 1 hits. (Courtesy of Kern County Museum.)

The most famous of Bakersfield's many honky-tonks was the Blackboard, where Bakersfield Sound pioneer Bill Woods and his Orange Blossom Playboys, pictured here in 1957, were the house band for 16 years. Buck Owens was Woods's lead guitar player until he formed his own Schoolhouse Playboys. Wider fame followed with Owens's first national hit in 1959; by 1964, he was a major star. (Courtesy of Kern County Museum.)

The Buckaroos, Buck Owens's band, enter through the backstage door at New York's Carnegie Hall on March 25, 1966. Owens was halfway through a string of 16 consecutive No. 1 singles when he was invited to appear at the famed venue, and was initially inclined to decline, wanting to avoid the embarrassment of unsold tickets. He reconsidered, the 2,700-seat venue sold out, and the band recorded a best-selling live album there. Unlike the majority of commercially successful Nashville performers who relied on the same small pool of studio musicians on their records, Buck Owens toured and recorded with his own band. Although a number of musicians would come in and out of the group over the years, most fans regard the classic era as the 1964–1966 lineup of guitarist/ fiddler Don Rich, bassist Doyle Holly, pedal steel guitarist Tom Brumley, and drummer Willie Cantu. These were the Buckaroos backing Owens when he headlined a multi-act show at the famed Manhattan venue in 1966. (Courtesy of Buck Owens Private Foundation.)

Merle Haggard, performing here live with his band the Strangers at a local radio station in 1970, became a hero of conservative America after his hit the previous year, "Okie from Muskogee," shot quickly to No. 1. Both Richard Nixon and George Wallace sought his endorsement during the 1972 presidential campaign; Haggard declined both. He later suggested the song was more parody than protest. (Courtesy of Kern County Museum.)

Four of the biggest stars of their era, Johnny Cash, Merle Haggard, Buck Owens, and Glen Campbell, pose here at a 1971 taping of the *Glen Campbell Goodtime Hour*, which aired from 1969 to 1972. Cash and Campbell had Bakersfield connections as well—Cash performed on Cousin Herb Henson's local live music television program, and Campbell lived and worked in Bakersfield before hitting it big in 1967. (Courtesy of Buck Owens Private Foundation.)

The Bakersfield arch, pictured here in all its glory in 1968, was the iconic symbol of the city to millions of travelers from the day it was erected over Union Avenue in 1949 until the California Department of Transportation ordered its removal in the late 1990s. Country music star Buck Owens built a replica arch just off Buck Owens Boulevard near his Crystal Palace dinner club in 1999. Owens has long been associated with the sign. It was built just two years before he moved to the city, and he used it as a backdrop for his memorable music video with singer Dwight Yoakam for their 1988 hit "The Streets of Bakersfield," a remake of an Owens song first recorded in 1972. The replica sign is, like the original, a yellow arch whose blue porcelain letters spell out the name of the city. It is supported by two towers whose design is inspired by the Beale Memorial Clock Tower. (Courtesy of Kern County Museum.)

This view of Union Avenue looking north toward Brundage Lane in 1949 underscores just how lively and energetic the gateway to Bakersfield was in the postwar years. Highway 99 once brought travelers right through the heart of Bakersfield, and there was something going on most every night, on or just off Union Avenue, in the African American clubs along nearby Lakeview Avenue or in the honky-tonks south of the city. (Courtesy of Kern County Museum.)

The Saddle and Sirloin was a popular night spot and upscale steakhouse on Union Avenue in the 1950s and 1960s. Here, Bakersfield police officers pretend to arrest the Tune Jesters, a musical comedy trio in March 1955. The Saddle and Sirloin closed in the 1970s and became the home of the Chateau Basque restaurant. It is now a church. (Courtesy of Kern County Museum.)

The intersection of Seventeenth and L Streets, seen here looking west in October 1963, was deserted for a reason. Two months earlier, the US 99 freeway bypass had started routing traffic around businesses that, for decades, feasted on travelers stopping for food, gas, and hotel rooms. Meanwhile, diners in outlying farm towns, suddenly situated along the new freeway route, dealt with packed parking lots. "Boom or bust," the newspaper called it. (Courtesy of Kern County Museum.)

Elsie Belle Urner, left, wife of Bakersfield Chrysler Plymouth dealer John T. Urner, affixes a "Kern County First" bumper sticker to her new Plymouth outside the Twenty-first Street store of her husband's brother David E. Urner, founder of Urner's Appliance Center, in 1971. Assisting her is Hilaire Ann Geraldine, spreading the Chamber of Commerce message: "Support Your Local Economy – It Supports You!" (Courtesy of Kern County Museum.)

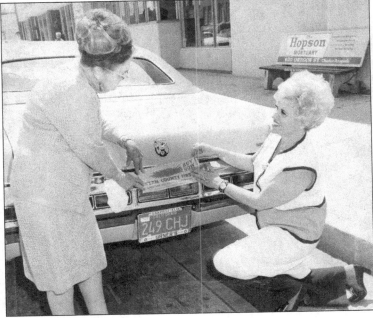

The Rancho Bakersfield Motel's prominent sign at Golden State Highway and F Street famously suggested to travelers, "Let's Eat." It beckoned for 40 years from what was once old US 99, which in those pre-freeway days was a gallery of neon-lit hotels, motels, restaurants, and nightclubs. It finally came down in the 1980s. (Courtesy of Kern County Museum.)

Customers belly up to the counter at the Friendly Café, home of the 12-egg omelet and eight-stack pancake breakfast, in May 1986. The Friendly, on North Chester Avenue, two blocks south of Buck Owens's recording studio in the former River Theater, closed in the early 1990s, but for decades it was the unofficial, early-morning annex to Trout's, one of Bakersfield's last-surviving honky-tonks. (Courtesy of Kern County Museum/Alan Ferguson.)

Congressman Bill Thomas, Republican of Bakersfield, speaks at a hearing alongside Congressman Leon Panetta, Democrat of Monterey, California, in December 1987. Thomas, a former Bakersfield College instructor, served in Congress from 1979 to 2007, finishing his tenure as chair of the House Ways and Means Committee. Panetta served from 1977 to 1993, and later was Pres. Bill Clinton's chief of staff. Later still, Panetta was director of the CIA. (Courtesy of Kern County Museum.)

Freda Hamilton enjoys a quiet lunch at Luigi's, the iconic Italian restaurant and delicatessen in Old Town Kern in East Bakersfield, in 1994. Luigi's, which was founded in 1910, has a remarkable collection of sports photographs and newspaper front pages from Bakersfield's glory days as a hotbed of sports dynasties and star athletes. (Courtesy of Kern County Museum/Henry Barrios.)

Civil rights activist Dolores Huerta, who moved to Bakersfield in 1965, is five feet, two inches tall and unassuming, but should not be mistaken for a pushover. In 1955, at age 25, she met César Chávez. In 1962, she and Chávez launched the organization that would become the United Farm Workers. She is the recipient of the 1998 Eleanor Roosevelt Award for Human Rights, the 2002 Puffin/Nation Prize for Creative Citizenship, the 2007 Community of Christ International Peace Award, and the 2013 Award for Greatest Public Service Benefiting the Disadvantaged. She has nine honorary doctorates and 31 acting credits. Her name is attached to two government holidays, at least seven public schools, and one asteroid. Pres. Barack Obama presented her with the 2012 Presidential Medal of Freedom. She has been arrested at least 22 times, and thanks to an encounter with the San Francisco Police Department in 1988, the year of this photograph, she no longer has a spleen. (Courtesy of Kern County Museum.)

The Great Dust Storm of December 19, 1977, shut down Bakersfield and beyond for three days, left five dead, and caused $34 million in damages. Gusts of 192 miles per hour stripped 25 million tons of soil from grazing land and row crops—enough to fill 1.6 million standard dump trucks. The force of the wind was so mighty that it sandblasted the paint off cars, pitted chrome bumpers, and knocked down billboards. The storm was caused by an unusually intense high-pressure system over the Rocky Mountains and a low-pressure system off the California coast. Nature moved to fill the vacuum, and Kern County was between the two pressure systems. The weather event—which was considered a "once in a lifetime" storm and named one of the top 15 weather events of the century by the National Weather Service—closed portions of all major freeways running through Kern County, stranded motorists, and knocked out power. "It looked like someone dropped an A-bomb on Bakersfield," pilot Dick Powers told the *Californian* newspaper. (Courtesy of Arvin-Edison Water Storage District.)

A now-infamous photograph of the great dust storm of 1977, taken from a twin-engine aircraft, shows what appears to be a wave of dirt reaching a height of nearly 5,000 feet moving over Arvin, a farm town south of Bakersfield. Winds reached 88 miles per hour in Arvin before the anemometer broke the first day of the storm, but the US Geological Survey estimated gusts of 192 miles per hour, according to the National Weather Service. By comparison, a Category 5 hurricane has sustained winds at 157 miles per hour and an EF-5 tornado generates wind speeds in excess of 261 miles per hour. LeRoy Schnell, the Arvin Edison water master dispatcher, told the *Californian* at the time that the wind was shooting gravel the size of peas through the air. In the immediate aftermath of the storm, residents noticed an eerie leftover—crickets. Millions of them were left behind by the deluge as a sort of parting gift. (Courtesy of Arvin-Edison Water Storage District.)

Two gargantuan sentries inviting freeway travelers to "Sun, Fun, Stay, Play" in Bakersfield stood at the north and south ends of the city for 17 years beginning in the late 1960s. Playful and cartoonish, the giant signs featured greeting-card poetry so memorable they developed a cult following. In 1965, the Heath Company of Los Angeles, hired to build the signs for $27,000 apiece, came up with the idea for the aspirin-shaped "balloon" signs, suggesting they contain the words "Eat, Rest, Gas, Play" and then "Eat, Rest, Swim, Play" before the final rhyme was selected. The identical structures—49 feet tall and 60 feet wide—were damaged by the great dust storm of 1977, and the ongoing maintenance costs began to irk business owners in the service district. Eventually, the signs came to be regarded as truck-stop tacky, and in 1983, they came down. The removal of the "Sun, Fun, Stay, Play" signs marked the end of Bakersfield's century-long era as a drive-through town and aptly represented a conclusion to the city's first 100 years. (Courtesy of Kern County Museum.)

DISCOVER THOUSANDS OF LOCAL HISTORY BOOKS FEATURING MILLIONS OF VINTAGE IMAGES

Arcadia Publishing, the leading local history publisher in the United States, is committed to making history accessible and meaningful through publishing books that celebrate and preserve the heritage of America's people and places.

Find more books like this at
www.arcadiapublishing.com

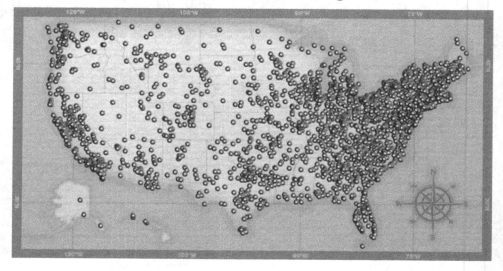

Search for your hometown history, your old stomping grounds, and even your favorite sports team.